Insects
and
Spiders

Beetles

Shane F McEvey
for the Australian Museum

This edition first published in 2002 in the United States of America by Chelsea House Publishers, a subsidiary of Haights Cross Communications.

Reprinted 2003

Chelsea House Publishers
1974 Sproul Road, Suite 400
Broomall, PA 19008-0914

The Chelsea House world wide web address is www.chelseahouse.com

Library of Congress Cataloging-in-Publication Data Applied for.

ISBN 0-7910-6600-2

First published in 2001 by
Macmillan Education Australia Pty Ltd
627 Chapel Street, South Yarra, Australia, 3141

Edited by Anna Fern
Text design by Nina Sanadze
Cover design by Nina Sanadze
Australian Museum Publishing Unit: Jennifer Saunders and Catherine Lowe
Australian Museum Series Editor: Deborah White

Printed in China

Acknowledgements
Our thanks to Martyn Robinson, Max Moulds and Margaret Humphrey for helpful discussion and comments.

The author and the publisher are grateful to the following for permission to reproduce copyright material:

Cover: A Christmas beetle, photo by Peter Marsack/Lochman transparencies.

Australian Museum/Nature Focus, p. 27; C. Andrew Henley/Nature Focus, pp. 5 (top and bottom), 18 (all), 19 (all), 28 (top); Carl Bento/Nature Focus, p. 26; Dennis Sarson/Lochman Transparencies, p. 25; Densey Clyne/Mantis Wildlife, pp. 8 (middle), 23 (right); Hans & Judy Beste/Lochman Transparencies, p. 9; Jiri Lochman/Lochman Transparencies, pp. 4, 8 (bottom), 10 (bottom), 11, 12 (top and bottom), 13 (top and middle), 15 (bottom), 16 (top and bottom), 17, 20 (bottom), 22 (top and middle), 23 (left), 24, 25 (top), 28 (bottom), 30; John Fields/Nature Focus, p. 13 (bottom); Michael Trenerry/Nature Focus, p. 15 (top); Mike Coupar/Nature Focus, p. 21; Pavel German/Nature Focus, pp. 6–7, 8 (top); Peter Marsack/Lochman Transparencies, pp. 10 (top), 22 (bottom); Phillip Griffin/Nature Focus, p. 14 (bottom); T. & P. Gardener/Nature Focus, p. 29; Trevor J. Hawkeswood/Nature Focus, p. 14 (top); Wade Hughes/Lochman Transparencies, pp. 14 (middle), 20 (top).

Contents

Glossary words

When a word is printed in **bold** you can look up its meaning in the Glossary on page 31.

What are beetles?

Beetles are insects. Insects belong to a large group of animals called invertebrates. An invertebrate is an animal with no backbone. Instead of having bones, beetles have a hard skin around the outside of their bodies that protects their soft insides.

Beetles have:
- six legs
- four wings
- two **antennae**
- two eyes
- a mouth
- many breathing holes on the sides of their bodies.

A paropsis beetle walks along a plant stem.

What makes beetles different from other insects?

Beetles have hard wing covers that meet on their backs in a straight line. These hard wing covers are the beetle's front wings and they are known as **elytra**. They protect the soft back wings that the beetle uses for flying.

A jewel beetle sits on a leaf. Its wing covers are closed, showing the colorful patterns on them.

Fascinating Fact

There are as many as 350,000 kinds of beetles throughout the world. More than 28,000 of these live in Australia.

This jewel beetle is about to fly away. It has opened its wing covers and extended the two soft wings underneath that it uses for flying.

Scientists have given a special name to all beetles. They are called **Coleoptera**.

Beetle bodies

The body of an adult beetle is divided into three segments. These segments are called the head, the **thorax** and the **abdomen**.

Adult beetles have hairs on their bodies and around their mouths. These hairs come in many different shapes and sizes. They can be long or short, thick or thin.

Did you know?

There are so many different kinds of beetles that only a very small number have common names. All the rest have a scientific name or no name at all.

abdomen

Abdomen

The abdomen is where:
- food is digested
- female beetles produce eggs
- male beetles produce **sperm**
- breathing holes are.

Thorax

On the thorax are:
* legs
* wings
* breathing holes.

Head

On the head are the:
* mouth
* **mandibles**
* antennae
* eyes.

thorax

head

This beetle is a lucanid beetle. Most lucanid beetles are active only at night, although some do feed on eucalypt or acacia leaves in daylight.

The head

On the head of an adult beetle are the mouth, eyes and antennae.

Mouth

Beetles have jaws called mandibles for biting, cutting, catching and chewing food. The mandibles can bite through wood and repel predators.

Eyes

Beetles have two compound eyes. A compound eye is made up of lots of tiny eyes packed together.

This is a longicorn beetle.

Antennae

Beetles use their antennae to touch and smell their environment. Antennae have a variety of shapes. They can be long, short, comb-like, feathery or saw-like.

Did you know?

Muscles are attached to the base of the antennae so beetles (and other insects) can move them around as they feel and sense their environment.

This beetle has very long and feathery antennae.

This paropsis beetle has very simple and straight antennae.

The thorax

On the thorax of an adult beetle are the legs, wings and breathing holes.

Legs

Beetles have six legs. They use their legs for walking, running, landing, digging and sometimes for swimming and jumping.

Wings

Beetles have two types of wings: a pair of hard wings on the outside called elytra and a pair of soft wings underneath. The elytra protect the soft wings that are used for flying. When flying, beetles use the strong muscles in their thorax to flap their soft wings while they hold their elytra open.

Breathing holes

Beetles breathe through tiny holes in the sides of their bodies called spiracles. These spiracles are protected by the elytra so the air a beetle breathes is not too dry, which would cause the beetle to dry out. Beetles do not breathe through their mouths.

This Christmas beetle is coming in to land on a plant stem. You can see its legs and wings. This beetle also has lots of hair on its body.

Did you know?

Most of the colorful markings on a beetle are on its elytra — the hard front wings.

elytra

wings

legs

Where do beetles live and what do they eat?

Beetles can live just about anywhere from the tropics to very cold places, from wet, humid rainforests to dry, hot deserts, from mountains to the coast and from dry, wind-swept plains to areas around fast-flowing rivers. Beetles will live in a certain place because that is where they find their food.

Beetle young (called **larvae**) eat lots of different kinds of food. They find their food using their sense of smell. What a beetle larva eats depends on what kind of beetle it is. Larvae can eat all the parts of a plant, including the wood, leaves, flowers, bark, roots, fruit and seeds. They can also eat all kinds of animals, both dead and alive. They also often eat other insects.

Adult Christmas beetles like to eat eucalyptus leaves.

Jewel beetles like to feed on plants. This jewel beetle is feeding from the flower of a native Australian plant.

Beetles in our houses

Beetles also live with us in our houses. You can find them in things we eat — cereals, grain, rice, oats and flour. You can find them in the things we wear — furs and cotton clothing. You can find them in our furniture like old cupboards and chests of drawers.

Have you ever had **weevils** in the flour in your kitchen? Have you seen the holes made by beetles that live in wood? Beetles that eat wood are a nuisance around the house, but they are useful in forests. There they help break down large pieces of wood and other plant material.

This weevil likes to live in dry places.

Fascinating Fact

There are more than 65,000 kinds of weevils in the world. Around 4,500 of these live in Australia. Some weevils like to live in areas where people live, and eat some of the things people eat.

Beetles that live in deserts and dry habitats

Some beetles can live in very hot and dry places, like deserts. Here are some of the beetles that can live in hot, dry places.

The bright green tiger beetle lives in dry places. Tiger beetles eat other insects. These beetles spend most of their time on the ground and they usually do not fly. They are mainly active at night. Being active at night when it is cooler is one of the ways beetles can live in hot, dry places.

Desert longicorn beetles live in the ground. Living in the ground is one of the ways beetles can live in hot, dry places.

Carcass beetles like to live in arid, dry regions like the center of Australia. Both the adults and larvae of these beetles eat dry animal remains. Carcass beetle larvae live in burrows in the ground under the dead animal's body.

Some dung beetles only eat wallaby dung. They cling to the hair around the **anus** of wallabies.

Dung beetles feed on animal **dung**. The adult beetles dig burrows beside or under the piles of dung. The beetles bury some of the dung in the burrows and then lay their eggs. They do this so the larvae that hatch from the eggs will have food to eat.

This dung beetle is starting to burrow into the sand.

These unusual beetles are called pie-dish beetles and they are only found in Australia. Pie-dish beetles live in hot, dry areas. Their elytra are fused together so they never open their wings. These are ground dwelling beetles that never fly.

13

Beetles that live in forests and wet habitats

Lots of beetles like to live in forests. Some beetles live under water. Here are some of the beetles that live in forests and wet places.

The color of this scarab beetle makes it hard to see when it is among the leaves, especially in wet rainforests. This helps it hide from predators such as birds.

Water beetles can live underwater. They can live in running streams or still ponds. Water beetles come to the surface to breathe. They can also trap air between their abdomen and elytra to take with them underwater. Water beetles have a smooth, boat-shaped body and flat, paddle-shaped hind legs which help them to swim. Water beetles feed on other water animals like tadpoles, dragonfly larvae and small fish. Adult water beetles can fly long distances to find isolated ponds to live in.

The ironbark beetle lives in northern New South Wales. It lives on or below the bark of eucalypt trees. It has a very hard thorax with 'shoulders' that extend forward. This protects the beetle's head as it pushes through narrow spaces under bark.

Fascinating Fact

Adult water beetles sometimes mistake shiny artificial surfaces, such as glass, for water.

Some beetles have bright metallic colors. This stag beetle is one of the most colorful. This is a lucanid beetle and it only lives in the rainforests of Far North Queensland. This beetle is usually active at night and likes to live in logs and stumps of trees.

Did you know?

Animals killed by cars and lying by the road are a good place to find beetles like carrion beetles. By eating and burying dead animals, beetles play an important role in nature.

Carrion beetles are sometimes big beetles. They can grow to four centimeters (1.5 inches). Their bodies are flattened, with lumps and bumps on their backs. Carrion beetles eat the rotting bodies of dead animals. Sometimes they bury parts of the dead bodies to feed their larvae.

How beetles communicate and explore their world

Beetles can get information about their environment in a number of ways. They can smell, feel, hear and see. While beetle eyes cannot see very well, their antennae can smell and feel extremely well. Some beetles can even make noises and glow in the dark!

All beetles have antennae that they use to find food or other beetles. They do this by sensing the special chemicals that food or other beetles release. They can also use their antennae to feel their way around the environment.

All beetles have compound eyes that can sense light and dark. Most beetles are active at night and are attracted to light. But some beetles, like jewel beetles, are active during bright, sunny days.

Beetles use their hairs to feel surfaces and objects. Beetles also feel sounds.

A group of plague soldier beetles on a gum tree. These beetles are exploring their world. They are feeding on the flowers and communicating with each other by touching with their antennae. Some of the beetles are mating.

A scarab beetle feeds on a flower. This beetle has a hairy body. Its hairs touch things and provide information about the environment.

16

How beetles communicate

Some beetles can make sounds. Most of the time it is the adult beetle that makes these noises but sometimes the larvae also make sounds. Beetles make sound by rubbing their legs together, usually their middle leg against their back leg.

Insects make sounds when they are seeking a mate. Sometimes the sounds they produce might ward off predators.

A click beetle sits on a plant. If a click beetle drops onto its back after a short while it will click its body and jerk into the air. You can feel this by holding the beetle between your fingers.

Did you know?

If you gently press a click beetle, it will try to click.

Fascinating Fact

Some click beetles and lampyrid beetles can glow in the dark. These beetles make light with a special organ in their abdomen. This light helps males and females find each other. Glowworms and fireflies are actually beetles that make light.

The life cycle of beetles

The whole life cycle of a beetle, from egg to adult, can take a few weeks or as long as a whole year. Some beetles only lay a few eggs while others lay many.

Beetles reproduce **sexually**. This means that a male and a female are needed to make new beetles. The male beetle provides sperm while the female beetle provides eggs. The eggs and sperm need to join together for a new beetle to start growing. Female beetles attract male beetles by releasing special smells or perfumes. When the male smells with his antennae, he follows the smell until he finds the female.

A male and female ladybird beetle mating.

Fascinating Fact

During the whole life cycle of a beetle, only the larva grows in size.

A new adult ladybird beetle sitting on its empty pupal skin.

Ladybird beetle pupae stay very still.

The adult beetle eventually emerges from the pupa. When the beetle becomes an adult, it does not grow any more. Some beetles live as adults for several years.

The female beetle lays eggs that hatch into larvae. Beetle eggs are always laid near or in food so that when they hatch, the larvae will have food to eat. Larvae are little worm-like creatures with legs, a head and a worm-like body. They do not look like adult beetles.

A female ladybird beetle next to her eggs.

The newly hatched larvae of ladybird beetles.

The larvae spend the first part of their lives eating and growing bigger. As they grow, their skin becomes very tight until it finally splits. This allows the larvae to grow even bigger in their new, larger skin. This is called **molting** and it happens several times during the life of a larva.

A ladybird beetle larva feeding on aphids.

The last molt of the larva is special. This is when the last larval skin splits open and the new larva is wearing a soft pupal skin underneath. As the soft pupal skin gradually hardens the larva becomes a **pupa**. In the **pupal stage** they stay very still and do not eat — they are turning into adult beetles.

This ladybird beetle larva is getting ready to become a pupa.

19

Larvae

Beetle larvae can be lots of different shapes and sizes depending on what kind of beetle larvae it is. Beetle larvae also live in lots of different places depending on what they eat.

Beetle larvae eat a lot and grow big quickly. Beetle larvae usually look very different from adult beetles.

Scarab beetle larvae live in the soil. When people dig them up, they sometimes mistake them for witchetty grubs. Witchetty grubs are the caterpillars of a moth and they don't curl up like these scarab beetle larvae.

When water beetle larvae are little, they usually live in water. As they grow bigger, they spend more time out of water. When they are in water, they eat small animals like fish and other insect larvae. Water beetle larvae have long hairs on their bodies.

Where beetle larvae live

Beetle larvae can live:
- in rotting wood
- in roots in the ground
- in living wood in trees
- on leaves
- in water
- in dead bodies
- in dry, stored food like flour
- in dung.

Longicorn beetle larvae eat wood as they tunnel through it. They often push piles of sawdust to the entrance to their holes. Beetle larvae that eat wood have hard, sharp mandibles that enable them to chew through hard wood.

Predators and defenses

Beetles are attacked or eaten by many different animals, such as birds, spiders, fish, lizards, mammals, frogs, wasps and other beetles. Animals that attack and eat other animals are called predators. Beetle larvae living in the ground have to protect themselves from **nematode worms** and **mites**. Beetles can also die from diseases and some types of **fungus**.

A small native Australian mammal called a ningaui attacks a large weevil.

A beetle larvae makes a good meal for this native Australian mulgara.

This scarab beetle larva is infested with mites. They are the small cream dots on the outside of the larva's body.

Protection against predators

Beetles protect themselves in a number of ways. Some beetles release a nasty chemical that can protect them from predators.

Beetles also defend themselves by hiding, pretending to be dead, running away, biting predators and coming out only at night.

Some beetles defend themselves by being very hard to see. Can you see the longicorn beetle in this photo?

Did you know?

The antennae of some longicorn beetles are long and saw-like and can cut your fingers if you handle them.

Some beetles, like these soldier beetles, occur in very large numbers. This helps protect the individual beetles from being eaten.

Weird and wonderful beetles

Welcome to the wonderful world of bizarre and extraordinary beetles!

Water beetles

A whirligig beetle lives on the surface of the water. When it is disturbed, it dives into the water and holds onto a weed. It carries a bubble of air down with it so it can breathe under water. The eyes of this beetle are divided into two sections so that it can see above and below the water at the same time.

Explosive beetles

The bombardier beetle has two chemicals in its abdomen. By themselves each chemical is harmless but when they are mixed together they explode. When the beetle is attacked it mixes the two chemicals together inside its body and then the mixture explodes out of the beetle's anus with a loud pop and a sizzle of steam. This beetle has a very hard anus to protect itself from the exploding chemicals.

Cleaning beetles

Some beetles called dermestids are used in museums for cleaning delicate skeletons of small mammals, reptiles and birds. The beetles eat dry animal flesh and leave the little skeletons intact. Dermestids can also be a pest in museums when they get in among the specimens and eat them.

Dermestids

Dung beetles

Some dung beetles will roll a ball of dung until it has a dry crust around it. This crust keeps the inside of the ball moist and fresh. The female dung beetle lays her egg inside the ball. When the egg hatches, the larvae eat the dung and grow inside the ball. Eventually a new adult dung beetle will break out of the crust.

The dung beetle uses its shovel-like head to cut out a piece of dung. It then shapes the dung into a ball.

Blistering beetles

When whip-lash rove beetles are irritated or crushed, they produce a fluid that can cause your skin to blister. The blister is like a thin line on the skin and often happens when you try to brush the insect off your skin with your hand. But the blister does not appear until one or two days after you have touched the beetle, so most people do not know that the beetle caused the blister.

Horned beetles

Some beetles have horns on their bodies. These may help protect them from being eaten by predators.

Fascinating Fact

Although there are tens of thousands of different kinds of weevils in Australia, they all have an easily recognizable shape.

This beetle is called a spiny weevil.

Collecting and identifying beetles

There are so many kinds of beetles that scientists are still discovering many new kinds. Some scientists spend years studying the beetles they have caught. If a scientist catches a beetle that is unknown, they name it and describe it so other scientists can study it too.

An example of each kind of beetle that has been found is kept in collections at museums. These collections are used by scientists who want to study the kinds of beetles that have already been found. Sometimes new beetles are discovered in museum collections long after they have been caught. Dead insects, like beetles that are in collections, do not rot like other dead animals because they are small and they dry quickly.

Some museums have tens of thousands of different kinds of beetles and millions of individual insect specimens in their collections.

When scientists collect beetles, they use special equipment. They sometimes use a light to attract beetles, nets and traps to catch them and jars and containers to put them in. Sometimes beetles are pinned to a piece of card and placed in collections. Each beetle collected will have a small label attached to it that has information about where the beetle was collected, when it was collected and who collected it.

Pinned beetles last longer because they do not bump around in containers, which could break bits off their bodies. The pin also gives the scientist something to hold onto when they want to examine the beetle.

How are beetles identified?

Beetles are identified by looking very carefully at their shape, size and color. If a beetle's shape, size and color is different to all other beetles that scientists already know, then this beetle is considered a new kind of beetle and is given a scientific name.

The shape of a beetle's antennae is important and helps distinguish different kinds of beetle.

What do scientists study about beetles?

After a beetle has been given a name, scientists study:
- where it lives
- what it eats
- when in its life cycle it is a pupa
- how often it molts
- what makes it turn into a pupa or an adult
- what its predators are
- what poisons or pollutants kill it or interfere with its normal life cycle.

Did you know?

Some of the beetles at the Australian Museum are over 100 years old but still look like ones collected this year.

A scientist studying beetles.

Ways to see beetles

How many different beetles can you observe around you?

- Hold an umbrella upside down under an **acacia** bush and beat the leaves with a stick. Look at the beetles that have fallen into the umbrella.
- Turn over logs and rocks to see what beetles live underneath. Make sure you turn them back afterwards so that you do not ruin the beetle's home.
- Look for weevils in food such as old pasta, flour, rice, muesli, bird-seed, rolled oats and cereals.
- Look for click beetles under outdoor lights on a summer night.
- Watch jewel beetles play dead when they are disturbed.

Watch ladybird beetles and their larvae feeding on **aphids**.

Watch how beetles lift their elytra before they start to fly.

How to keep beetles

Many beetles can be kept as pets. Some large beetles live for many years. To keep a beetle as a pet you will need:

- **A container for the beetle to live in**
  Fish tanks, biscuit tins and shoe boxes are good containers to keep beetles in.

- **Things for the beetles to hide under**
  There should be wood and leaves in the container for the beetle to hide under. Make sure any heavy pieces of wood don't roll around and squash your new pet.

- **Food for the beetle to eat**
  Some beetles will be happy with rolled oats, bird seed and fish food and other beetles would prefer fresh leaves or dead insects. It is important to know what type of food your beetle likes to eat and feed it that food. Try to see what the beetle is eating before you catch it.

- **Moist tissue paper**
  Keep the air in the container moist by keeping a little piece of damp tissue paper in the container. When the paper dries out, you will need to wet it again.

Rhinoceros beetles can be kept as pets and like to eat apple cores.

Remember!

- You will need to clean the container and supply fresh food about once a week.
- Do not forget to feed your beetle the type of food it likes to eat.
- If you give your beetle what it likes to eat and how much light and water it likes, it probably will not try to escape.
- Make sure you keep the container out of the sun. It can get too hot.

Good beetles to keep include:
- carab beetles that feed on small insects
- weevils that live in dry food like flour and oats
- mealworms that feed on dry oats and flour
- rhinoceros beetles that will eat apple cores.

Beetles quiz

1   On what part of its body is a beetle's mandibles?

2   How many wings do beetles have?

3   What is the special name scientists have given all beetles?

4   What are a beetle's hard front wings called?

5   How many legs do beetles have?

6   Does a beetle have its elytra open or closed when it is flying?

7   Where do beetles lay their eggs?

8   Do beetles grow any bigger after they emerge from their pupa?

9   Do beetle larvae look the same as adult beetles?

10   Why do some beetles play dead and drop to the ground when disturbed?

11   How many different kinds of beetles are there in Australia?

12   Do beetles have hairs on their bodies?

13   When elytra are closed how do they close along the beetle's back?

14   How do beetles make sound?

15   What is the name used to describe the way a larva sheds its tight skin as it grows?

Jewel beetles feeding on native flowers.

Check your answers on page 32.

Glossary

abdomen	The rear section of the body of an animal.
acacia	The scientific name for a group of plants that includes wattles and mulga.
antennae	The two 'feelers' on an insect's head that are used to feel and smell. (Antennae = more than one antenna.)
anus	The hole in the rear of an animal.
aphids	Small insects that suck sap from plants. Aphids are bugs.
Coleoptera	The scientific name for beetles.
dung	Animal droppings.
elytra	The hardened front wings of a beetle that cover the soft hind wings.
fungus	Mushrooms and toadstools are kinds of fungus. Some fungi grow flat on or under the skin of animals. (Fungi = more than one fungus.)
larvae	Caterpillars, grubs and maggots are kinds of larvae. In the life cycle of an insect the larval stage is after the egg stage and before the pupal stage. Larvae hatch out of eggs, grow and then turn into pupae. (Larvae = more than one larva.)
mandibles	Hard structures for biting.
mites	Small spider-like animals that are not insects.
molting	When an animal sheds its entire skin it molts. The process is called molting.
nematode worm	A kind of worm (not an earthworm) that is usually pointed at the ends, shiny, wriggles in loops, and is often smaller than 1 cm (0.39 in) long.
pupae	In the life cycle of an insect (like moths, beetles and flies), larvae turn into pupae. Adults later emerge from pupae. A cocoon is a shell around a pupa. (Pupae = more than one pupa.)
pupal stage	A stage in the life cycle of an insect when the insect is a pupa.
sexual reproduction	When a male and female living thing combine to make more living things.
sperm	The male reproductive cell.
thorax	The middle section of an animal's body.
weevils	The name for a very large group of beetles that scientists call the Curculionidae.

Index

Answers to quiz

1 head 2 four 3 Coleoptera 4 elytra or wing covers 5 six 6 open 7 near the larval food 8 no 9 no 10 to avoid being eaten by predators 11 28,000 12 yes 13 in a straight line 14 by rubbing their legs together 15 moulting.